WHALEMEN AND WHALESHIPS OF MAINE

1. What this Kennebec County Maine painter lacked in skill he more than made up for in vitality and detail. A stricken sperm whale has shattered a whaleboat with his flukes, scattering men and gear in all directions: paddles, line tub, bucket and irons are visible. Undaunted, the second whaleboat's boatsteerer has darted his iron. From a private collection.

WHALEMEN AND WHALESHIPS OF MAINE

A PUBLICATION OF
THE MARINE RESEARCH SOCIETY
OF BATH, MAINE

Kenneth R. Martin

HARPSWELL PRESS
Brunswick, Maine

Harpswell Press
Simpsons Point Road
Brunswick, Maine 04011

Library of Congress catalog card number: 74-28773
International Standard Book Number:

0-88448-030-5 paperbound
0-88448-031-3 hardbound

Designed by Wladislaw Finne
FIRST EDITION
Printed in the United States of America

CONTENTS

INTRODUCTION

Maine's involvement in whaling can be thought of as occurring in three spheres. First, shore whaling off the state's coastline appears to have taken place from the time of the earliest settlements up to the twentieth century. Secondly, a few Maine ports could not resist the success of the whaling centers in Massachusetts, Connecticut and New York so they outfitted their own whalers for voyages. These ran from dismal failures to moderate successes but never captured local interest sufficiently to allow whaling to be sustained beyond a few years in the state. Many Maine men went a-whaling in vessels from other ports in those days as well.

Lastly, since Maine's most significant contribution to the country's maritime history has always been shipbuilding, quite naturally Maine-built vessels were used in whaling from other areas. Among these were merchant ships whose use as fast freight carriers was limited after years of service, but which were still good for many more years afloat under the less rigorous demands of whaling in the South Pacific. In addition, vessels were built on contract in Maine specifically as whalers, particularly for New Bedford-area interests in the latter half of the nineteenth century.

For *Whalemen and Whaleships of Maine*, Dr. Martin has combed an impressive array of sources, some very obscure, to bring together for the first time a volume on this state's involvement in a business which never ceases to fascinate millions.

8 Possessing a rare combination of diligent scholarship and lucid, entertaining exposition. Dr. Martin has given us a book which will appeal to the student of maritime history as well as to the casual reader.

When this book was written, Dr. Martin hailed from the unlikely port of Slippery Rock, Pennsylvania, where he was a history professor who was interested both in whaling and in Maine. Since that time he has become registered out of Sharon, Massachusetts, where he is Director of the prestigious Kendall Whaling Museum, attesting to his credentials as author of this fine volume.

Dr. Charles E. Burden
Bath Marine Museum
Bath, Maine

ILLUSTRATIONS

ACKNOWLEDGMENTS

The author wishes to thank the Chicago Historical Society for permission to publish a painting from John F. Martin, "Journal of a Voyage in the Whale Ship Lucy Ann of Wilmington, Del.," in its manuscript collection. The Bath Marine Museum, Bath, Maine; the Kendall Whaling Museum, Sharon, Massachusetts; and The Whaling Museum, New Bedford, Massachusetts, are to be thanked for contributions from their pictorial collections.

Much of this material has appeared in significantly different form and detail in the following articles:

"The Folk Art of Maine's Whalemen," *The Quarterdeck*, II, 11 (Spring–Summer, 1972), pp. 10–11.

"Voyages of Maine Whaleships," *Down East: The Magazine of Maine*, XVIII, 8 (May, 1972), pp. 46–49, 75ff.

"The Bath-built Steam Whalers," *Down East: The Magazine of Maine*, XIX, 5 (January, 1973), pp. 44–47, 78ff.

"Spirit Voices and a Bath Whaler," *Down East: The Magazine of Maine*, XX, 4 (November, 1973), pp. 42–45, 65.

Many students of whaling and maritime history afforded me enthusiastic assistance. Special thanks are due Mr. Robert B. Applebee of Stockton Springs, Maine; the late Mr. Charles F. Batchelder, of Milton, Massachusetts; the late Mr. M. V. Brewington, Director Emeritus of the Kendall Whaling Museum, Sharon, Massachusetts; Charles Burden, M.D., President of the

Marine Research Society of Bath, Maine; the late Mr. Reginald B. Hegarty, Curator of the Melville Whaling Room of the New Bedford, Massachusetts, Free Public Library; Mrs. Barbara E. Johnson, of Princeton, New Jersey; Mr. Nathan Lipfert, of the Bath Marine Museum, Bath, Maine; Mr. Philip F. Purrington, Senior Curator of The Whaling Museum, New Bedford, Massachusetts; and Professor Stuart C. Sherman, John Hay Librarian at the Brown University Library, Providence, Rhode Island.

Thanks are also due to Mr. James Renckly for special photography.

2. This enormous sperm whale's tooth is a souvenir of a memorable 90-barrel bull taken off New Zealand by the whaleship Wiscasset. *The incised drawing of the* Wiscasset *on the tooth provides important information about the vessel's appearance. Courtesy of the Kendall Whaling Museum, Sharon, Massachusetts.*

WHALESHIPS FROM MAINE PORTS

American whaling was a Yankee enterprise, and during the peak years of whaling, New England vessels—hundreds of them—pursued the whale on every ocean of the world. In the mid-nineteenth century there were more American whaleships than the combined whaling fleets of all other nations, and throughout the period of this remarkable enterprise, the majority sailed from Massachusetts, southern New England, and Long Island ports.

Maine's distinctive contributions to whaling were of consequence in many ways. The discovery of the first known American whaling activity at Pemaquid even predates the birth of the entire industry.

In the seventeenth century whales were quite plentiful along the Maine coast.[1] The original Pemaquid settlement, begun prior to 1625 as a fishing and trading station, was also engaged in shore whaling—confirmed by archaeologists' recent discovery there of a whale bone fragment and a cutting spade. Like the Indians before them, these early settlers undoubtedly pursued whales spotted from the shore, killing them and towing their bodies ashore for rendering into oil.[2] But this kind of whaling was secondary to Pemaquid's systematized fishing enterprise.[3] Moreover, whales quickly learn to avoid dangerous shorelines, so whaling, like old Pemaquid itself, disappeared.

By the mid-eighteenth century Nantucketers had developed deep-water whaling into a highly complex and exclusive business. Although the industry suffered greatly during the Revolution and War of 1812, it revived and grew in the early nineteenth century; in the optimistic years before the American Civil War, more and more vessels went to sea from Yankee ports. While Maine did not develop many substantial blue water whaling ventures of her own, her chief contributions to the industry were her seamen—independent, confident men who shipped on the lengthy whaling voyages; the renowned Maine shipyards, which turned out many superior ships for the trade; and the business enterprises of a few coastal towns; all of which add a fascinating chapter to Maine maritime history.

In eastern Maine, some periodic shore whaling was carried out, most notably at Prospect Harbor and on Mt. Desert Island. As in the old Pemaquid days, whales spotted from shore were chased in small boats, harpooned and towed to shore-built try-works. U.S. Fisheries records indicate that Prospect Harbor probably took whales as early as 1810 and as late as 1870.[4] At nearby Winter Harbor, in 1845, a Captain Bickford sailed his schooner, the *Huzza*, along that part of the shore looking for whales and took eight.[5] A generation later, the Boothbay Harbor steamer *Fannie Sprague* whaled along the Maine shore from 1881 to 1885.[6] This sporadic shoreside industry resembled similar small town efforts in Massachusetts and on Long Island, New York. As on Long Island, however, Maine people were still shore whaling, at least at Vinalhaven, in the early twentieth century.[7] Whales can still occasionally be seen along the Maine coast, and their awesome size should impress anyone with the resourcefulness of downeast fishermen.

The rugged, unusually well-built whaleships themselves, built in Maine shipyards, represent downeasters' efforts to break into the highly lucrative deep-sea whaling industry. Then between 1834 and 1843, Maine businessmen formed whaling companies and sent ships to nearly every ocean in every latitude in search of whales. What brought downeasters, at long last, into the business of blue water whaling? A better question might be: what had kept Maine people out of the business for so long? We will never know the whole story; but the answers to both questions may lie in the nature of whaling, the times themselves, and the Maine man's temperament.

Southern New England and Long Island had managed to virtually monopolize whaling in America by 1830 because this was a highly specialized and complicated business, calling on whole communities to concentrate their efforts and resources for a long time, collectively, to squeeze profits from a dangerous, unpredictable enterprise. Those who were good at it, like the

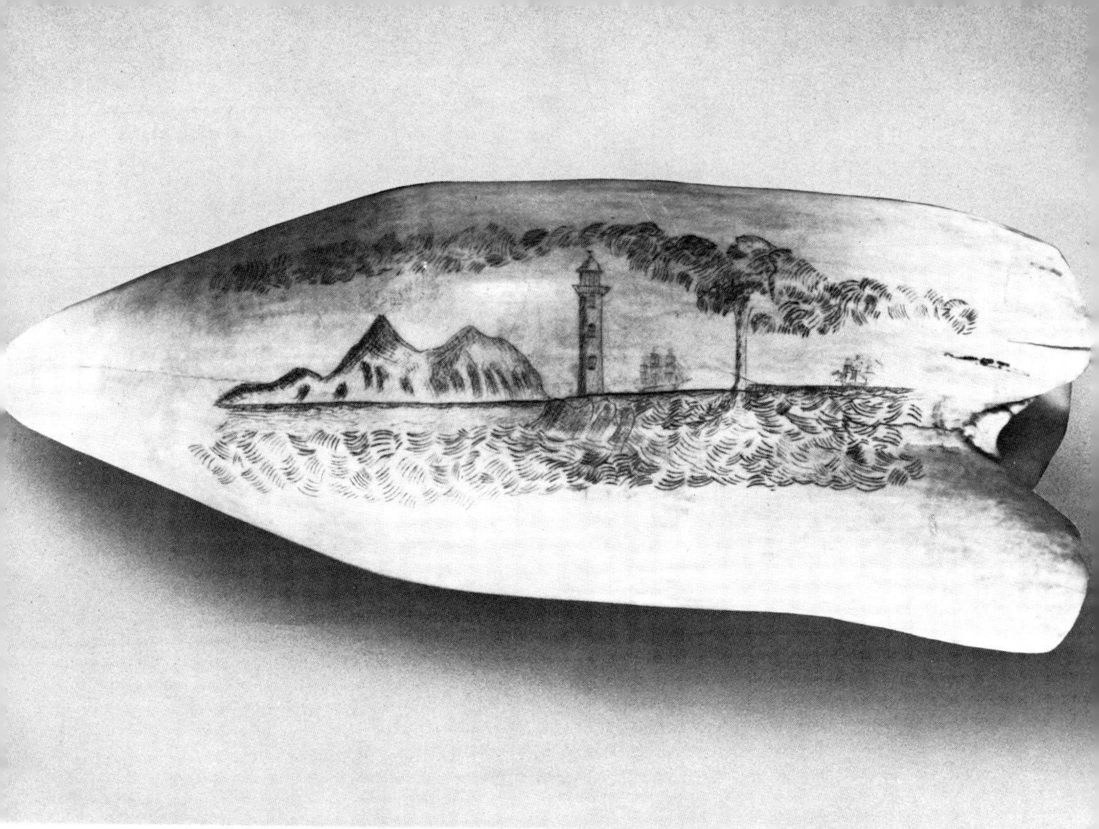

3. *A fanciful seascape adorns the reverse of the nine-inch* Wiscasset *tooth. Perhaps the scene expresses the wishful anticipation of a homeward-bound whaleman. Courtesy of the Kendall Whaling Museum, Sharon, Massachusetts.*

 Nantucketers, fared handsomely, but the demands were terrific and often produced no quick gain. It has been suggested, and it may be true, that Maine people lacked the cooperative collective spirit necessary for group specialization.[8] The downeaster, according to some, was an independent spirit and was temperamentally unable to mortgage himself to community endeavor. Maine ports, though, were busy with other commerce, and Maine's remoteness from the whaling centers also probably kept whaling from fully catching on.

The early 1830s was a period of great optimism and economic expansion in America. If Nantucketers and New Bedfordites could get rich in whaling, why not the people of Poughkeepsie, or Wilmington, or Portland? Under the impression that a little capital and a little luck would bring success in whaling, cities throughout America's northeast decided to try their hand. Portland, Wiscasset, Bath, and Bucksport were among these newcomers to whaling. Almost all failed within a decade or so. Like the rest, Maine people found out that whaling was hardly a get-rich-quick enterprise.

Portland was the first Maine town to enter the booming whaling business. Oddly, little is known of this venture, perhaps because her single whaleship, the *Science*, seemed a small item compared to Portland's other maritime trade. It does appear that Portland merchants, like those of many other towns, simply hired trained whalemen from southern New England and fitted out a ship, under the correct impression that America's expanding economy of the 1830s would insure high demand for oil, high prices, and consequently great profits.

The *Science* was a ship of 388 tons, built in 1827 at Newbury, Massachusetts.[9] Her skipper was a Captain Whippey, who may have been the former captain of the whaler *Martha* of Nantucket.[10] Quite likely the *Science*'s officers were also seasoned whalemen, and much of her gear was probably manufactured in a big-time whaling port. Whaling demanded all kinds of specialized gear unnecessary in other vessels, and whalers carried great quantities of spares and repair equipment. On the *Science*'s main deck, for instance, stood a brick tryworks with tremendous cauldrons for boiling whale blubber into oil. Starboard cutting platforms and massive blocks for hoisting blubber strips and whales' heads aboard were other necessities. Moreover, there were the whaleboats themselves. The *Science* carried four thirty-foot whaleboats, three of which could be lowered when whales were raised. Replacements for these fragile craft were stowed aft atop a shelter on the main deck. Each whaleboat had to be equipped with hunting gear: harpoons and lances, tubs of line, a cutting spade, a knife and hatchet, water cask, food and candles, boat hook, compass, signal waifs, bailing bucket, mast and sail, steering oar, rudder, and oars for a crew of six. Cutting-in a whale required such gear as mincing knives, cutting spades, lengths of chain.[11]

In January 1834, Captain Whippey took the *Science* out.[12] She was cleared for the Pacific Ocean.[13]

Once at sea, the green crew of a whaleship began a grueling shakedown. Divided into whaleboat crews, the "greenies" had to practice lowering their boats from davits and pulling together. Whaleboats, because of their lightness, could all but fly over the water when well-manned; but chasing whales could be a hard day's work, often involving miles of back-breaking rowing. Lightness gave a whaleboat its speed, but such a craft could easily be dashed to pieces by the immense flukes of a whale. Customarily a whaleboat pulled swiftly and silently alongside an unsuspecting whale, whereupon the boatsteerer darted his irons from the bow. Almost always, the surprised whale reacted by convulsively thrashing his flukes. Sperm whales often reacted angrily to the darted iron, charging the whaleboat head-on or sounding and rising mightily from the depths to smash or chew up the whaleboat—unless the crew could somehow maneuver the craft out of the whale's path. This required steady nerves and great teamwork. In any case, when the whale had tired, the whaleboats had to approach for the kill. The boat's officer then darted his lance into the whale's vitals, churning the blade until the agonized beast, drowning in his own blood, threw a column of thick, hot blood out his spouthole and finally turned "fin out." All this called for few prior skills, but necessitated much practice and toil once the whaleship was on her way. Ahead lay that first real chase and the panicky moment when a green crew slipped alongside the black immensity of a whale.

What was that first moment of truth like? Although no record of the *Science* crew's first encounter has apparently survived, we can turn to a contemporary description by another Maine whaleman. Here is David Densmore's recollection of his introduction to whaling at close quarters, as a hand aboard the Nantucket ship *Peru* in 1833. The scene was the South Atlantic,

Yours truly.
D. C. Dinsmore

 where the *Peru* had lowered her boats. The whales sounded:

> . . . We had some time to wait for their reappearance; the captain standing upon the stern sheets, while the boatsteerer [harpooner] was doing the same thing on the bow of the boat, both looking out for them. While this was going on I was cogitating in my mind what a whale looked like, how big his eyes were; for I imagined that, if they were proportioned like other animals, their eyes must be as large as a good-sized cart-wheel. While thinking of this and other things, the captain jumped down into the boat, and seized the steering oar, saying in a suppressed tone, "Pull three and stern two," meaning to pull ahead on the three-oared side and back-water on the other or two-oared side; by doing which the boat would swing almost square round, without going ahead. I looked in every direction, my eyes as big as demijohns. Directly I heard a roaring sound. The whale had come up, and the sound I heard was the noise he made in spouting. The old man had seen him under water, when he gave the above order . . . which manoeuvre brought the head of the boat towards him, and when he broke water he was within striking distance; and the boatsteerer let fly first one harpoon and then the other. Then came the order, "Stern all," meaning, back-water with all

4. This engraving of David C. Densmore by J. S. Conant appeared in Densmore's The Halo: An Autobiography. *Densmore is depicted here about 1875, a generation after the demise of the mysterious Bath Whaling Company. Renckly photograph.*

> oars. The whale making his appearance so unexpectedly, and so near us, we, that is the green-horns, got confused, at least I did; for I didn't know whether I was on my head or heels, and never knew whether I pulled ahead, or backed-water, or whether I did either. Now came the killing process. The whale was a large one, and seemed decidedly opposed to being interfered with in that unceremonious matter. He kicked and thrashed round furiously. . . . Finding he couldn't get clear by kicking with his enormous flukes, or tail, he commenced running to windward at a furious rate, dragging the two boats after him. . . . Then the officer uses his lance . . . with this they pierce the vitals, which sets the whale to bleeding. Nearly an hour elapses before they turn up, as whalemen say when the whale is dead. Our whale was soon checked in his running-away propensity. . . . In a shorter time than it takes to write this paragraph, Mr. Whale was spouting blood; and in thirty minutes more he was a lifeless mass of blubber.[14]

Densmore's account successfully evokes the fresh excitement of the chase, but such encounters as these quickly became old hat to experienced whalemen.

Captain Whippey took the *Science* to put in a season on the Offshore Grounds, west of Chile and Peru; a season in the South Pacific; one on the Japan grounds; and another right whaling off New Zealand. The progress of the *Science* can be measured by reports of the vessels who spoke her. Almost two years later, off Peru, she had taken only 500 barrels. Her luck changed though, off Japan in 1836. In September she was spotted during a rare period of pleasant weather by the New Bedford ship *Canton*. Both the vessels' crews had their hands full, as reported by a *Canton* whaleman, who gives us a rare, first hand glimpse of the *Science*'s activities:

> The first part of This Day we loard the boats for . . . whales. But, painful to remark that after many maneuvers one boat got along side of a large whale, but by dint of luck or devil didn't fasten noways . . . & consequently he cleared himself of the tackling soon he was off & about the same time when the above transactions were a taking place, the boats of the above mentioned Ship whilst on a tight chace came near our Ship. A one of these boats came on board & reported the Ship, to be the *Science* of Portland with 1200 bbs [in] 32 months.[15]

For several more days the *Canton* and *Science* sailed in company, chasing whales together and exchanging visits.[16] By November 1836, the *Science* had stowed down 1,500 barrels.[17]

The long, hard cruise had taken its toll, however: one man, Amos Burnham, had been lost overboard, and several others had deserted, probably in the Hawaiian Islands.[18] Contrary to the expectations of many aspiring foremast hands, life aboard whalers was so fraught with tedium that desertions were commonplace.[19]

The *Science*'s quarry was the sperm whale, whose oil was quite valuable but often difficult to come by. So it was that the *Science*, to fill her hold, had to make a long voyage. More than four years passed before the vessel was ready to return home.

Captain Whippey put into Talcahuano, Chile, sold some oil, refitted, and at last sailed for Portland, for the ship was long overdue.[20] Captain Whippey's take of 2,100 barrels, unloaded at Portland, represented at best a mildly successful voyage. The *Science* had been out four years and four months, and the unusual duration of the cruise undoubtedly cut deeply into profits.[21]

Five months later, on September 5, 1838, the *Science* cleared again for the Pacific.[22] This time she was out after right whales—a type now endangered—as well as sperm, and headed for Bay of Islands, New Zealand.[23] Right whale oil was in less demand, but right whales were easier to find and catch. Moreover, their upper jaws yielded hundreds of pounds of hornlike baleen. Baleen, called "whalebone" in the trade, serves as a kind of sieve in which the swimming right whale traps krill, the small crustacea on which it feeds. Baleen could be removed from the whale's jaw in strips several feet long, and had special application for such items as carriage whips and umbrella frames. It brought several cents a pound in the 1830s.[24] When spoken in late December, the *Science* had already taken three whales.[25] Captain Whippey was back in less than three years with a full ship: 3,100 barrels of right whale oil; only 300 sperm.

With this the Portland owners decided to call it quits on

5. John Patten, one of Bath's leading shipbuilders, standing before his North Street home, where David Densmore claims to have proposed a Bath Whaling Company. The house is still a Bath landmark. Courtesy of the Bath Marine Museum, Bath, Maine.

 whaling.[26] By 1841 the economy had changed; the United States was in a severe depression which ended quite a few small-time whaling ventures conceived in happier years. Under these circumstances, Portland could scarcely compete with Nantucket or New Bedford. The *Science* was later sold at New Bedford, where she continued as a whaler.[27]

Shortly after the *Science* had sailed on the first Maine whaling voyage, a downeast whaler left Wiscasset. The activities of this vessel, the ship *Wiscasset*, are better remembered than those of the *Science* and have been retold many times, perhaps because the community treated whaling with a bit more municipal enthusiasm.

Once a prominent trading center, Wiscasset had been for some years in decline. Then in 1833, fired by dreams of cashing in on the rising prices and growing demand for whale oil, and perhaps inspired by such New York river towns as Hudson and Poughkeepsie—communities that had formed lucrative stock corporations engaged in whaling—Wiscasset entered the trade. Citizens of the town pooled resources and formed the Wiscasset Whale Fishing Company, with William M. Boyd as president.[28] Whaling enthusiasm was high, and even before the *Wiscasset* was completed she made her way under jury rig down the Damariscotta River, through the narrows of Townsend Gut, and up the Sheepscot to Parsons' Wharf, where she was fitted out for a long whaling voyage.[29] As with Portland, the townsmen and merchants of Wiscasset early learned how costly the whaling business could be. Much of the *Wiscasset*'s gear had to be transshipped from New Bedford.[30]

 Only about twenty shareholders held stock in the company, but clearly the whole town was intrigued by the 380-ton vessel fitting out at Parsons' Wharf. In honor of the new enterprise, the wharf's name was changed to Whaleship Wharf.[31] Enough spirited hands were recruited locally to work the vessel to the Azores, where she could ship the remainder of her crew.[32]

Like the *Science*, the *Wiscasset* was a sperm whaler, a vessel which sought sperm oil, so prized as an illuminant, and spermaceti, a waxy material from the whale's head, which made the best candles. The sperm whale yielded less oil than the right whale, but its quality was purer. Sperm oil often brought three times the price of common whale.[33] Since whaling crews worked on a profit-sharing basis, it is easy to see how Maine boys might eagerly sign on for such a venture.

The *Wiscasset*'s skipper for her maiden voyage was Captain Richard Macy of Nantucket, a seasoned veteran, formerly of the Nantucket whaleship *Harvest*.[34] It was Macy's intention to put in a season on the Offshore Grounds, then head for New Zealand.[35] She would be gone more than three years.

The *Wiscasset* sailed down the Sheepscot River on May 13, 1834, painted with the conventional row of simulated gun-ports along each side.[36] She was not built for speed, but speed was one thing whalers did not need. She rounded the Horn, losing her rudder during the stormy passage.[37] Shipshape again, she disappeared into the Pacific. Her voyage is largely a mystery, but a few facts can be documented. The whaling was fairly good, for, when spoken north of New Zealand by the *Christopher Mitchell* of Nantucket, she had taken 1,350 barrels in

6. A tight spot sperm whaling on the Wilmington Whaling Company's ship Lucy Ann, *built at Bath, Maine in 1830. John F. Martin, a* Lucy Ann *crewmember 1841–1844, painted this watercolor depicting an actual incident in his journal of the voyage. Courtesy of the Chicago Historical Society Library.*

less than two years.[38] In these waters, the *Wiscasset*'s crew took the largest whale of the entire voyage: a bull sperm, master of a harem of cows. The *Wiscasset*'s boatmen had never seen such a whale; his blubber stowed down nearly ninety barrels.[39] The men saved the jawbone of this colossal bull, extracting the huge ivory teeth for souvenirs.[40] Several of the teeth have survived as scrimshaw, including one on which a crew member incised a scene of the whale's capture. This, the only picture of the *Wiscasset* available for public view, shows the vessel hauled aback as her boats draw near the whale. The tooth is currently on display at the Kendall Whaling Museum in Sharon, Massachusetts.

With more than 2,800 barrels aboard by 1837, the *Wiscasset* had been lucky indeed. One crewmember was lost—a routine matter aboard a whaleship—but no other mishap marred the voyage.[41] When she at last had rounded the Horn again and made her way north in the Atlantic, word of her success had long since preceded her to Wiscasset. The entire community jubilantly awaited her; and when, on September 10, 1837, she tied up at Whaleship Wharf, her return signalled a noisy, festive celebration.[42] The souvenir jawbone was hung up at wharfside as a permanent tribute to the voyage.[43] In one voyage, the *Wiscasset* paid for herself, the crew's shares, and returned a dividend to her stockholders![44]

Once again the *Wiscasset* sailed on a Pacific voyage, commanded this time by Captain Seth Horton of New Bedford. The vessel made her way around the Horn, putting in a season off New Zealand and another among the islands of the South Pacific. This voyage was not as spectacular as the first, but it was nonetheless successful.

However, the *Wiscasset*'s second voyage was a troubled one. Some of the crew deserted at Tahiti, and Captain Horton had to recruit again at Talcahuano, Chile. Also, the cabin boy, perhaps overcome by the mean hardships of whaling, set a fire aboard ship, forcing the *Wiscasset* to put back to Tahiti.[45] In spite of this, Captain Horton, out for sperm and right whales, filled the *Wiscasset* with oil and baleen. The first eleven months, during which the ship crossed the Pacific to New Zealand, were absolutely luckless.[46] But in the next two month, 850 barrels were taken, and in another year she had stowed down 1,700.[47] Whaling alternately in the waters off the Society Islands and New Zealand, Horton finally took 2,800 barrels of sperm and right whale oil in the Pacific before heading home in the spring of 1841. In July of 1841 the *Wiscasset* again returned heavy-laden: 1,200 barrels of right whale, 900 of sperm, and the cash from 600 barrels sold at Bahia on the homeward leg.[48]

Twice the Wiscasset Whale Fishing Company had been lucky. It was, however, pushing its luck. Expenses were higher than in the great whaling ports, and oil prices fluctuated so much the company could ill afford a bad cruise. Thus it elected to discontinue whaling.[49] The Wiscasset Whale Fishing Company had been incorporated by the Maine legislature at a maximum capital of 150,000 dollars.[50] Such a charter was a case of overambition, in view of Wiscasset's failure to acquire more than a single vessel. It seems likely, then, that after 1839 the onset of depression persuaded the company to risk no further ventures.

The *Wiscasset* was sold to Sag Harbor buyers, from which port she continued to whale.[51] In later days she served as a transatlantic packet for the New Line of New York, and on one passage she brought the young immigrant Andrew Carnegie to America.[52]

Although her days as a whaling port were few, Wiscasset recalls them fondly. On top of the town's municipal building perches a weathervane of recent vintage by Gil Whitman, patterned on the *Wiscasset*.[53]

In 1841, Bath tried whaling. This enterprise is all but forgotten, yet it is the most unusual one in Maine's brief whaling history. The origin of the Bath Whaling Company is tied up with the strange career of David Densmore, whose recollection of his first whaling encounter appears above. In later life, Densmore, a mystic, became a Boston faith-healer and occult newspaper editor. In this capacity he wrote and published his autobiography, *The Halo*, in which the former seaman outlined the obscure story of Bath whaling.[54] His reminiscences are supported by other unpublished sources.[55]

Densmore, a native of China, Maine, made three Nantucket whaling voyages between 1832 and 1840, rising from green hand to second mate. During these years, he survived some tough scrapes, and came to believe his good judgment in critical times was inspired by the guiding voices of departed souls.[56] In 1841, at twenty-eight, between voyages and almost broke, Densmore was staying in a Bath flophouse when a "voice" inspired him to raise funds for a Bath whaleship by soliciting subscriptions.

7. *The bark* Hannah Brewer, *built in Robbinston in 1850, typifies Maine-built merchant vessels that found their way into the whale fishery. Shown here in a gouache by a Neapolitan port painter during her merchant days, the* Hannah Brewer *became a New London, Connecticut whaler in 1852. Courtesy of the Kendall Whaling Museum, Sharon, Massachusetts.*

To start the fund, Densmore sought out influential Bath businessmen. He proceeded on the advice of his innkeeper to the house of the celebrated shipbuilder John Patten.[57] Patten agreed to contribute only if Densmore could raise the full amount by subscription—about 20,000 dollars. Apparently, Patten's support, limited though it was, did the trick: in a short while, Densmore had gotten promises from an impressive group of Bath businessmen that covered the price of a vessel.[58] Meeting in Samuel Stinson's counting room, the subscribers chartered the Bath Whaling Company. The thirty-eight charter members then purchased the brig *Massasoit*, a 206-ton merchantman built in Plymouth, Massachusetts in 1824–1825, and had her altered to a bark.[59] The fitting out of the *Massasoit* was accomplished with frugality. It coincided precisely with Portland's abandonment of whaling, and when the *Science*'s gear went up for auction it was bought at half price and used to equip the *Massasoit*. The new whaler's captain was to be none other than David Densmore himself. According to Densmore, his was a case of instant success:

> . . . it was unanimously voted that I should take charge of fitting out the expedition, and command the ship when ready for sea. The affair had been gotten up so quietly, that hardly any outside of those engaged in it knew any thing about it; and scarcely one, even among the subscribers, until after the first meeting, knew my name . . .[60]

The advice of Densmore's "voice" had produced big dividends—at least for the moment.

Because Bath ship craftsmen were unfamiliar with turning out essential whaling craft, the departure of the *Massasoit* was delayed a number of months.[61] She sailed with an inebriated, motley bunch:

> My crew all told, including officers, amounted to thirty-four men, mostly foreigners. They represented all grades of society, some having held important positions among their fellows; but dissipation, one way and another, caused them to fall from their high estate. . . . As a whole, the crew was an average one for a whaler. A large per cent of them were "green hands," who, as soon as we got to sea, were sick: and, the old sailors being drunk, the chance for taking in or making sail were not very flattering.[62]

Several of the crew were from Densmore's native China.

Three weeks out of Bath, off the Cape Verdes, Densmore believed a mutiny was in the offing, led by forecastle grousers afraid of rounding the Horn. Seizing the initiative and his pistol, Densmore assembled the crew and made a "stump speech," warning the crew of the dangers of insubordination and threatening due consequences for ringleaders. If the crew had had any mutinous schemes, they apparently dropped them then and there.[63]

Off Brazil, two men took sick. The *Massasoit* made for St. Catherine's to seek help for them, but both died nonetheless. In May 1842, the vessel was off the River Plate and still empty.[64] Because of the seasonal Cape Horn weather she stayed in the

 Atlantic. In July 1842 she made Fayal with 120 barrels of sperm; and, a year after her departure, off Brazil, she still had taken only 230 barrels.[65] Six months later the *Massasoit* had taken 400 barrels of sperm oil; she sold some of this, and headed home in February 1843 with only 300 barrels.[66] Homeward bound she searched the Gulf of Mexico and the Hatteras whaling grounds and stowed down 100 more.[67]

Densmore did not find the *Massasoit* a very good sailer. After leaving Bath, she had to be hastily rerigged as a ship for proper management; and once in a bad gale off the River Plate Densmore found "the ship laboring very hard, her upper works twisting and working like a raft. The decks leaked, and at one time I had serious apprehensions whether she would weather it or not."[68]

The whaling may have been none too good, but Densmore's "voices" saw him through more than one tight spot. One of these occurred off Cape Cod. Advised by a passing vessel to sail offshore to avoid a northeaster, Densmore heeded his "voice's" call to tack inshore. Instead of a northeaster, the coast was struck by a "smoking southwester." The *Massasoit* handily made Seguin Light and the Kennebec without a mishap, but the skipper who had told Densmore to keep offshore was lost with his vessel.[69]

The hectic voyage came to an anticlimactic end. Making the Kennebec on March 1, 1843, the *Massasoit* got upriver almost to Bath when she went onto a rock.[70] She had to be lightened by the removal of her oil, tryworks, and topmasts before she could be floated. When she finally made Bath, whatever interest there

8. The whaling bark William Baylies, *outward bound on her maiden voyage. In this oil by C. S. Raleigh of New Bedford, the* Baylies *has passed Dumpling Lighthouse and crew members are bidding goodbye to wellwishers aboard the tug. Courtesy of the Kendall Whaling Museum, Sharon, Massachusetts.*

had been for whaling had cooled, although Densmore claims the voyage still finished up in the black.[71] Some of this may have been insurance payments and a profit on the resale of the vessel, however. The *Massasoit* returned to the merchant service.[72] Years later, she became a whaler again, with Mattapoisett, Massachusetts, her home port.[73]

Densmore's whaling career ended at Bath, but many interesting years and occupations lay ahead for him before his decision to become a professional occultist and writer. His whaling memories, set down a generation after the fact, suffer from possible lapses of memory about chronology and small details, but his unusual story appears to be essentially accurate. One can only wonder what the reactions of Bath's practical businessmen would have been had they realized the Bath Whaling Company's existence was inspired by a disembodied "voice"!

While the *Massasoit* was fitting out at Bath in 1841, another curious whaling voyage began from Bucksport, on the Penobscot River. In April of 1841 the schooner *Warwick* sailed from that town on an Atlantic whaling cruise. She did not get far. Bound for the Azores, the *Warwick* was battered by foul weather and had to return home. Here is her misfortune as reported by a Portland newspaper:

> Sch. Warwick . . . on a whaling voyage, put back to Bucksport, 26 inst. in distress, having been knocked down on the 10th . . . while lying to in a heavy gale: had decks swept of everything except the tryworks, lost three boats, caboose, and house binnacle, booby hatch, oars, whaling craft and mainsail. She had

> a crew of 15 men, nearly all of whom were washed overboard but saved as if by a miracle, one man having caught a piece of rigging, by which he regained the vessel, and the others being washed back upon her decks by the returning wave.[74]

The *Warwick* put to sea again in June.[75] Her master, Captain Grogin, is otherwise unknown in whaling annals. He may, like Densmore, have been a local downeaster, but this is speculation. The *Warwick* was a "plum pudd'ner," a small vessel which sailed for a single season in the North Atlantic. Such voyages were the specialty of Cape Cod whalemen, who customarily sailed to and from the Azores, crossing the Western Grounds in search of sperm whales.

Outward bound, the *Warwick* took blackfish.[76] These small whales, easily taken, yield good quality oil and were often chased by outward bound vessels to give whaleboat crews practice. The *Warwick* sold her blackfish oil in the Azores and returned to the nearby Western Grounds.[77] Her voyage, though, was a failure, for when she came home in April 1842, she had taken only 110 barrels of sperm.[78] Hard luck sealed the fate of Bucksport's whaling venture.

So it was that, within a few years, Maine seamen and investors learned that whaling was hard and expensive work, best handled by those who had the organization and special skills for such a trade. Maine's handful of whaleships, though, are a priceless, colorful sidelight in her history, symbolic of the adventuresome spirit of downeasters.

Did any other ships sail to the whaling grounds from Maine

 ports? It is quite possible. For example, manuscript records mention a brig *Beulah* of Bangor, which may have whaled in 1838, but nothing else is known of her.[79] There is another curious mystery involving a journal kept aboard the whaleship *Elbe* of Hamburg, Germany, 1845–1848, which refers to the *Elbe* as a Portland vessel.[80] The keeper of that journal, George F. Neil, was an American mate, and there may have been many other Yankees aboard. But why Neil occasionally refers to the *Elbe* as a Portland vessel is a puzzle. Was she partly owned by Portland people? Does Neil perhaps refer to Portland, England? Did the European turmoil of 1848 perhaps affect her registry? The *Elbe* journal is one of those curiosities that demonstrates the enigma a document can create. Perhaps other forgotten Maine whaleships will be discovered in the future.

9. *The busy Bath shipyard of Goss, Sawyer and Packard in June 1876. The brand-new bark afloat in the Kennebec is the whaler* John and Winthrop. *Courtesy of the Bath Marine Museum, Bath, Maine.*

MAINE SHIPS FOR THE AMERICAN WHALING FLEET

The indirect contributions of Maine—her shipyards and men—were more important to whaling than the state's short-lived whaling enterprises. Maine-built vessels played a part in the expansion of Yankee whaling. They also were of critical importance in the era of decline, for they helped postpone the demise of the industry.

While whaling was on the increase, the demand for well-built, durable vessels grew. Shipyards far from New Bedford built an occasional whaleship. Maine yards supplied a good share of these, since they were solid and low-cost. Some celebrated whaling vessels hailed from downeast.

Often, though, a merchant vessel would be bought and refitted as a whaler. A representative example is the *Lucy Ann*, built in Bath in 1830.[81] A three-masted, 309-ton merchantman, the *Lucy Ann* was bought by the Wilmington Whaling

 Company of Wilmington, Delaware, another town caught up in the whaling craze of the 1830s.[82] This vessel whaled out of Wilmington for a decade, during which she worked whaling grounds in every corner of the globe. When Delaware whaling folded, the *Lucy Ann* whaled from Greenport, New York, finally being condemned at Rio in 1850.[83]

A more long-lived example of Maine vessels converted to whalers is the bark *Tamerlane*, built at Wiscasset, which logged twenty-eight years of service before entering the whale fishery at New Bedford in 1842.[84] Forty years later she was still at it, until she went on a reef at Hilo and was lost with most of her hands.[85] Another Maine-built whaler, the little bark *Janet* of Westport, Massachusetts, gained a bizarre notoriety. In June 1849, while whaling in the Pacific, one of the *Janet's* boats was upset by a sperm whale. The boat crew, including the captain, managed to right the empty boat, but they were lost to the *Janet*, drifting for twenty-one days without provisions.[86] As crewmen succumbed, their bodies provided food for the survivors. In desperation, lots were drawn by survivors to select a crew member to die for the nourishment of the others. Only the captain and one other survived the ordeal; these two were later rescued from Cocos Island by another whaleship.[87]

Often, Maine-built vessels were extensively renovated for whaling. The Bath-built brig *Jasper*, for instance, was rebuilt as a bark before whaling out of New Bedford.[88] All in all, almost two hundred American whaleships originated in Maine yards, a great many coming from the Kennebec.[89] These ranged from small schooners to the sophisticated steam barks of the late nineteenth century, built especially for whaling.

Whaling was basically a conservative business, tradition-bound and parsimonious. Although steam vessels came in at the end of the 1870s, most whaleships continued to be built along traditional lines, and wooden sailing barks predominated. Between 1876 and 1878, for instance, the Goss, Sawyer and Packard yard at Bath was kept busy by the famous New Bedford owner William Lewis, whose firm commissioned the barks *John and Winthrop, James Allen, Josephine, Lucretia* and *Mabel.*[90] And although Lewis pioneered the use of steam whalers, he nonetheless continued to invest in sailing vessels. In 1886, the New England Shipbuilding Company, heir to Goss, Sawyer and Packard, turned out the beautiful whaling bark *William Baylies* for Lewis.[91] The *Baylies* and the *Lucretia,* incidentally, were later converted to steam.[92] All of these vessels except the *John and Winthrop* were lost, the last disaster being that of the *Josephine* in 1919.[93] So it was that Maine played a notably productive part during the whaling industry's last effective years.

Whaling reached its peak about 1850. After the American Civil War, the industry never recovered its prewar vigor. During the Civil War, the American whaling fleet had been reduced by Confederate raiders; dozens of other whaleships had been bought by the U.S. government for the "Stone Fleet" and sunk off Charleston to block that harbor. Whale oil, meanwhile, had

Cruising off Santa-Cruz or Nitendi Ids

Lat 10 33 S Long 165 45 E Monday May 2 1887 fine with light wind from E Ship in between Guerte & Linekula Ids. Plenty Natives alongside Harried weather all night. Squalls of wind & Rain from N and Ships close in bight between Cape Trevenin & Carteret Point and came very near going ashore

Lat 10 32 Long 165 44 Tuesday 3rd fine wind light from E.N.E. Landed for [illegible] Plenty Porpoises. Took in Sail near Linekula

Lat 10 35 Long 165 50 Wednesday 4 A.M rain squally P.M fine. dead calm. Repair Lower Fore Topsail. Saw plenty Porpoises & two Swordfish.

Lat 10 24 Long 165 57 Thursday 5 Very fine with a nice breeze from E by S. Cruising E of "Linekula". Saw the "Swallow Group". 4 Canoes came off from "Santa Cruz"

Friday 6 fine. Dead calm till 12-30 then light airs from E.S.E. and raised Whales 10 or 12 miles E of Linekula. Lowered 4 Boats and L.B struck one in the fin joint and bent the point of the Iron as if it had been against an Anvil. W.B also [illegible] in the Small. a long dart. both Irons drawed two Boats came onboard at 4 P.M the others chased till dark without success. Plenty Whales Saw a lot of large Cows 30 or 40 in all two large ones seen.

Saturday 7 fine with light breeze from N.N.W. Raised Whales at 7 a.m S.E Linekula 12 miles. [illegible] 8 O C and L.B & S.B got a nice [illegible] Whale at 1 P.M and cut in at 6-15, one turned up close in to the Beach. carried all Sail while cutting to work off shore. Plenty Natives on board Started Tryworks.

25 bls S B 25 bls L B

Sunday 8 Very nice weather with a good breeze from W.N.W. Raised Whales at 9 a.m close in to the Beach [illegible] Swallow Bay. Lowered 3 Boats and W.B [illegible] S.B [illegible] Cow both turned [illegible] the Beach. the L.B had to tow off. being to near to go with the Ship. had them fluked at 1-15 and cut in at 7-15 kept the Try works going during taking the Whales and cutting. Plenty Whales. Took the last Whale alongside directly N of Cape Byron.

25 bls S B 25 W.B 25

10-42 -166-13 Some rain Squalls in night

10. This page from the logbook of the Bath-built bark John and Winthrop *of New Bedford reveals by its whale stamps a busy week's work in May 1887: six whales struck and four taken. The use of whale stamps in logbooks was widespread in the fishery. Courtesy of the Kendall Whaling Museum, Sharon, Massachusetts.*

lost much of its market as the world turned to kerosene. For a while, the reduced demand and reduced numbers of whaleships balanced, so there was still money to be made. All the same, Yankee owners and their captains became more pinchpenny than ever. The romance of whaling was wearing off; it was now well-known that whatever excitement whaling offered a young man was overpowered by the long voyage, its brutality and boredom, and the meager financial rewards. Whaling interests had to rely more and more on deceptive recruiting or shipping crews of sullen recalcitrants, many from waterfront flophouses. Other crewmen could almost always be shipped at the Azores, Cape Verdes, or even the Arctic itself, and hard-nosed Yankees continued resolutely to squeeze every penny out of whaling.

The fishery got a new lease on life when, as fashion changed, Victorian corset- and dressmakers sought baleen for stays. The price of baleen rose to dizzy heights, making it worthwhile to hunt down the right whale and its Arctic cousin, the bowhead.

Bowheads yield hundreds of pounds of baleen and as much as 150 barrels of oil each. The bowhead was comparatively plentiful in the western Arctic during its spring and fall migrations.[94] The 1850s and 1860s saw more and more whaleships heading for the western Arctic in the spring, where vessels would drift northward with breaking ice for a summer season in the Arctic, often getting as far as Point Barrow. Many such whalers stayed icebound through the long Arctic winter in order to maximize the next spring and summer season. Arctic whaling required extreme endurance, and it is small wonder few willing whalemen could be found. Moreover, the bowhead itself was a frustrating quarry. Sometimes seen in great numbers, it might all but vanish for a season. The bowhead is an almost harmless animal, but bowhead whaling could be very dangerous, since when struck these whales frequently headed for ice floes, taking the whaleboats on a collision course with the ice as they sounded.[95] This hazard could be overcome somewhat by the newly-developed bomb-lance, a device which discharged an explosive projectile in the whale's vitals seconds after impact, often killing him outright.

The combined hazards of Arctic whaling were great indeed. The whaleman's greatest worry was the perfidious Arctic weather, especially shifting winds that could send a crushing mass of drift ice against the hull of a motionless whaler. In spite of very heavy losses of whalers in the 1870s, the Arctic continued to attract entrepreneurs. Another factor was the completion of transcontinental rail facilities, which allowed whalers to operate out of San Francisco, discharging oil and baleen for freighting east and returning annually to the Arctic.

To reduce Arctic risks, some American whalemen turned to a new, specially designed vessel, one that could minimize the Arctic threat. Such planning produced a group of Bath-built steam barks that did much to sustain whaling in the late nineteenth century. These vessels, modelled on the British Dundee whaleships which had proven themselves in Greenland waters during the 1870s, inspired William Lewis to construct an American steam whaler for his New Bedford firm.[96]

The *Mary and Helen*, a steam whaler constructed in 1879 for

Lewis, was something of a celebrity, and the firm of Goss, Sawyer and Packard spared no expense in turning out this elegant vessel. Framed in white oak, the 420-ton *Mary and Helen* was 139 feet long, and was carefully reinforced to withstand great pressure on her hull. Bark-rigged, she could make eight or nine knots on her steam power alone—in a clear channel.[97] In Arctic waters, the steam power would enable her to make headway or get under way in the ice. She also sported steam-driven deck equipment to ease the crushing strain of hoisting blubber and the heads of whales aboard. The *Mary and Helen* was a luxurious vessel inside and out, and the captain's quarters were nothing less than opulent. As a refinement, C.A.L. Simpson, a Bath figurehead carver, embellished her sharp bow with a figurehead of Mary Haskell, and her round stern with a quarterboard carving of Helen Haskell, both of whom were shareholders in the new vessel.[98]

After an enthusiastic sendoff from Bath, the *Mary and Helen* left New Bedford in September 1879.[99] Fitted out at San Francisco with whaling gear, she headed north to try her luck. She returned full after one season, her voyage a bonanza.[100] William Lewis lost no time in commissioning more steam barks at Bath. In July 1880, his second, the *Belvedere*, left the Kennebec. In 1881, a third, the *North Star*, was built, and the *Mary and Helen II* begun.[101] A curious fate, however, awaited the first *Mary and Helen*. After her single whaling voyage, she was sold to the U.S. government, to track the wrecked Arctic explorer ship *Jeannette*. She never whaled again, and was lost when she caught fire in 1881.[102]

12. *Under full sail and flying the red-and-white house flag of William Lewis, the steam whaler* North Star *sails off on her first and last voyage. Built at Bath in 1881, she was lost the next year at Point Barrow, Alaska. The painting is by C. S. Raleigh. Courtesy of the Kendall Whaling Museum, Sharon, Massachusetts.*

now reversed, sold the brand-new *Mary and Helen II* to the Pacific Steam Whaling Company; renamed the *Beluga*, she almost outlived the industry. Her end, however, was curious. Converted during World War I into a sailing vessel, she was lost on a commercial voyage.[112] Not many steam barks were so long-lived. The *Lucretia* went down off Herald Island in 1889.[113] The *William Lewis* was on her third voyage in 1891 when she too was wrecked off Point Barrow.[114]

Perhaps the worst of all these calamities was the one that befell the *Navarch* in 1897. In July of that year the *Navarch* became hopelessly trapped in moving ice. Her crew trekked for another vessel, the cutter *Bear*.[115] Captain Joseph Whiteside, his wife and first mate crossed the ice successfully and found safety, but many crewmembers, lost, wandered desperately in search of help. Sixteen died by drowning, freezing and suicide, but a few survivors managed to get aboard another ship in the area, the *Thrasher*.[116] When the *Navarch* froze in again, several crewmembers who had stayed with her disembarked to safety, leaving behind two others who refused to evacuate even then; these holdouts soon drifted north to oblivion when the *Navarch* went adrift once more. Several other Arctic whaleships were lost that season, their crews rescued by the *Bear*.[117]

By the turn of the century, both oil and whalebone prices had plummeted. Fewer and fewer vessels made the trip to Point Barrow, but even at this late date a Maine-built vessel entered the whaling trade. At San Francisco, in 1904, the old Bath-built steam barkentine *Morning Star* got a new lease on life as the whaler *Herman*.[118] The *Morning Star* had done service supplying mission outposts in the Pacific islands; now this veteran joined other steam barks in the Arctic. But the days of Arctic whaling were all but over.

In 1908, with twenty voyages behind her, the *William Baylies*'s time came. Her skipper, Hartson Bodfish, a career Arctic whaleman, had been seduced by whaling twenty-eight years before, when at New Bedford he beheld the *Belvedere* set off on her maiden voyage.[119] Bodfish and the *Baylies* were an impressive team, but in May 1908, they met their match at Plover Bay, Siberia. While whaling there the *Baylies* became sandwiched between drifting ice and land ice. Her bottom was torn away, and the crushed vessel began to settle at once. This was an emergency all Arctic whalemen were prepared for, and the *Baylies* was quickly abandoned without incident, the crew making its way to another nearby whaler. As they departed, the *Baylies*'s stern settled and she sank rapidly.[120]

The *Belvedere*'s good fortune held for a long time. She whaled out of Seattle well into the twentieth century; then, in 1919, she was lost off Siberia.[121] The *Thrasher*, converted to a merchant schooner, the *Kamchatka*, caught fire in the Aleutians and was lost in 1921.[122] The *Herman*, ex-*Morning Star*, had one last role to play. In 1923 she disappeared into Mexican waters as the motor ship *Chapultepec*.[123]

In the long run, luck proved to be of prime importance in Arctic whaling. The Bath steam barks, well-built though they were, were built for a chancy business and inevitably ran out of luck, some sooner, others later. The vessels helped revive interest in whaling for a time, however. Their demise signalled the end of the industry, and with their passing came the finish of Arctic whaling.

13. *Another William Lewis steam whaler, the* Mary and Helen II, *painted by C. S. Raleigh, who depicted her crewmen investigating a derelict. Sold by Lewis and later rebuilt as the* Beluga, *this vessel made many Arctic whaling voyages. Courtesy of The Whaling Museum, New Bedford, Massachusetts.*

14. *The ill-fated steam bark* William Lewis *whaling in the Arctic, as painted by C. S. Raleigh. Though she was as ruggedly constructed as her sister ships, she fell prey to Arctic perils on her third voyage in 1891. Courtesy of The Whaling Museum, New Bedford, Massachusetts.*

MAINE MEN IN AMERICAN WHALING

In the great days of whaling it was a cliché that whaleships could easily ship green crews made up of venturesome boys from the farms of Pennsylvania, New York, Maine and Vermont. Maine boasted a number of hardy sailors, many drawn south to the bustle of New Bedford. Whaleship masters seldom asked many questions when they recruited and runaways could easily ship out. Whaling's novel appeal seems to have lured Maine men in large numbers. Shipping lists give evidence of Maine men in Yankee whaling crews at least up to about the time of the American Civil War. The crew list of the New Bedford whaleship *Addison*, 1857–1859, for example, mentions three Calais, Maine hands in the initial crew, two of whom deserted. Another Maine man was signed on at Honolulu.[124]

Some of these downeasters are well known in

 whaling literature. The most affecting character in J. Ross Browne's famous memoir, *Etchings of a Whaling Cruise* (1846), is a hapless Maine boy whose naive goodness invites abuse by his officers.[125]

In contrast to this seaman is George Harrison, a participant in the notorious *Junior* mutiny of 1858. Harrison came from the Maine woods, and found his woodsman's knowledge handy while hiding from authorities in the Australian bush.[126] Another typical of the many downeasters who made good in the fishery is Captain Amos A. Chase, an East Limington youth who rose to captain in eleven years of whaling.[127]

Although the names of most Maine whalemen are today forgotten, their vast numbers are evident in the prevalence of scrimshaw throughout the state. Clearly whalemen appreciated the uniqueness of their calling, for detailed whaling scenes often appear on the whalebone, baleen and whales' teeth they decorated for mementos. Many of these were brought back to Maine farms by sadder-but-wiser men whose escapism had been cured by a single whaling voyage.

Today whaling folk art has become so collectible that genuine pieces are rare indeed. Important pieces are still turning up in old Maine houses, however, and from time to time a modest item of scrimshaw can be found at a summer flea market. There seems to be a disproportionately large amount of scrimshaw in Maine. Perhaps Maine seamen and their descendants appreciated better than others the special drama of their calling. Whatever the case, this folk art inspires interest in a vanished and adventurous industry and remains a charming testament of the downeast men who went whaling.

Sobering indicators of whaling's grimmer side exist too, especially in newspapers of the period. Here are some typical items from Portland papers of the 1840s:

> The whale ship *Thule,* Capt. Coffin, on the morn of the 11th Sept. struck on a shoal WNW from the Booby Shoals . . . . the mate's boat got into the breakers on Cato's Bank, and it is feared was lost, with all on board. Two of the crew in the mate's boat, we notice . . . are from this state, viz: Alonzo O. Hatch, of Kennebec, and John Huntley; name of town not stated.[128]
>
> Killed by a whale . . . off River Platte, D. MacDonald, of Eastport, aged about 18 years . . . on board whaling brig *Troy,* of Bristol Rhode Island . . . . Killed by a whale, June 18, Mr. Abraham Butts (or Stubbs) of Portland, Me., aged 24, seaman, belonging to the whaling bark *Theo. Chase* of Westport Massachusetts.[129]
>
> On board whaling barque *Hope,* Mar. 18, 1846, John D. Clark, of Augusta, Me., boatsteerer. He was killed by a whale.
>
> Drowned by the upsetting of a boat, Feb'y 13, 1846 . . . Ethan Forbush, of Sanford, Me., one of the crew of the ship *Adeline Gibbs,* of Fairhaven, aged 21.[130]

An 1852 item in the *Eastern Argus* reports the loss of the Fairhaven schooner *Lively,* a whaler built at Chelsea, Maine. The

15. The Navarch, *photographed at the Kelley and Spear docks at Bath in 1892, shortly after her completion. The* Navarch's *loss in 1897 was one of the most calamitous ordeals of Arctic whaling. Courtesy of the Bath Museum, Bath, Maine.*

names of the missing include three Hallowell men, including the skipper, Job Pierce; a man from Vassalboro; and another from Eastport.[131]

The dangers of seeking and killing the gigantic mammals took their toll, but whalemen succumbed even more to illness and shipboard accidents. Illness on a whaler could mean a lingering, inevitable death. Such was the case with Alvan Robinson of South Norridgewock, Maine, who died aboard the ship *Eliza Adams* of New Bedford in 1850. His passing was briefly noted by a shipmate:

> Alvan Robinson died Oct. 22^{d} about 6 oclock in the morning after about 2 weeks sickness with the Disinterry, his fathers name is Increase Robinson, South Norridgwock, Maine.[132]

Maine's contributions to whaling of her ships and her men unquestionably enhanced the dynamism of American whaling. If her own business ventures from Maine ports were sporadic and short-lived, they yet rounded out downeasters' participation in every activity of this enormously colorful trading enterprise, and added another unique aspect to the story of American whale fishery.

NOTES

1. Captain John Smith, *Travels and Works of Captain John Smith, President of Virginia and Admiral of New England*, ed. by Edward Arber (2 vols., Edinburgh: John Grant, 1910), p. 187; see also Ida S. Proper, *Monhegan, Cradle of New England* (Portland, Maine: Southworth Press, 1930), pp. 123–124; and Robert P. Tristram Coffin, *Kennebec; Cradle of Americans* (New York: Rinehart & Company, Inc., 1937), pp. 204–206.

2. For an eyewitness account of Kennebec Indian shore whaling methods, see James Rosier, *A True Relation of the Voyage of Captaine George Waymouth, 1605*, in Henry S. Burrage, ed. *Early English and French Voyages, Chiefly from Hakluyt* (New York: Charles Scribner's Sons, 1906), p. 392.

3. Helen Camp, *Pemaquid Lost and Found* (n.p., 1967), pp. 3, 12–13.

4. A. Howard Clark, "History and Present Condition of the Fishery," in G. Brown Goode et. al., *The Fisheries and Fishery Industries of the United States: Section V, History and Methods of the Fisheries* (3 vols., Washington, D.C., Government Printing Office, 1887), II, p. 40.

5. *Ibid.*, p. 41.

6. *Bulletin of the United States Fish Commission, Volume V, for 1885* (Washington, D.C., Government Printing Office, 1885), pp. 169, 207.

7. Ivan E. Calderwood, *Sequel . . . Days of Uncle Dave's Fish House* (Rockland, Maine: Courier Gazette, Inc., 1972), pp. 89–94.

8. E. Lee Dorsett, "The Maine Whalers," *Down East: The Magazine of Maine*, II, 3 (April 1956), p. 14.

9. National Archives Project, *Ship Registers of New Bedford, Massachusetts* (3 vols., Boston: National Archives Project, 1940), I, 290, no. 2855.

10. Federal Writers Project of the Works Progress Administration of Massachusetts, *Whaling Masters*, American Guide Series (New Bedford: Old Dartmouth Historical Society, 1938), p. 289.

11. A good discussion of a whaleboat's gear may be found in Captain Charles M. Scammon, *The Marine Mammals of the North-western Coast of North America Described and Illustrated Together with an Account of the American Whale-Fishery* (1874; rpt. New York: Dover Publications, Inc., 1958), pp. 224–225.

12. Alexander Starbuck, *History of the American Whale Fishery from its Earliest Inception to the Year 1876* (1878; rpt. 2 vols., New York: Argosy-Antiquarian Ltd., 1964), pp. 314–315.

13. "Ship Science of Portland," in Dennis Wood, Abstracts of Whaling Voyages from the United States 1831–1873 (manuscript, Melville Whaling Room, Free Public Library, New Bedford, Massachusetts).

14. David C. Densmore, *The Halo: An Autobiography* (Boston: Voice of Angels Publishing House, 1876), pp. 82–83.

15. For my own gratification and satisfaction I write . . . A Journal of the Most Noted Proceedings and Transactions . . . which Generaly Arises on a Voyage a Whaling to the Pacific (manuscript journal kept aboard ship *Canton* of New Bedford, 1834–1838, no. 43, Kendall Whaling Museum, Sharon, Massachusetts), 20 September 1836.

16. Journal Kept aboard the Ship *Canton* of New Bedford, (manuscript, 1835–1838, no. 252, Kendall Whaling Museum, Sharon, Massachusetts), 20 September 1836; see also Charles C. Norton, Journal of a Voyage to the Pacific Ocean in the Ship *Canton* (manuscript journal, 1834–1838, no. 355, Kendall Whaling Museum, Sharon, Massachusetts), 19 September 1836.

17. Wood, "Ship Science of Portland."

18. *Eastern Argus* (Portland, Maine), 9 March 1837.

19. Elmo Paul Hohman, *The American Whaleman* (1928; rpt. New York: Library Editions, Ltd., 1970), pp. 62–64.

20. *Daily Evening Advertiser* (Portland, Maine), 9 March 1836.

21. Starbuck, pp. 314–315.

22. *Eastern Argus*, 5 September 1838.

23. Wood, "Ship Science of Portland."

24. Walter S. Tower, *A History of the American Whale Fishery*, University of Pennsylvania Publications on Political Economy and Public Law, no. 20 (Philadelphia: Published for the University of Pennsylvania, 1907), p. 128.

25. Phylander Gifford, Remarks on Board Ship Addison (manuscript journal, 1838–1841, no. 5, Kendall Whaling Museum, Sharon, Massachusetts), 30 December 1839.

26. Densmore, p. 145.

27. Starbuck, pp. 410–411.

28. George Wharton Rice, *The Shipping Days of Old Boothbay from the Revolution to the World War, with Mention of Adjacent Towns* (Boothbay Harbor, Maine [Portland, Maine]: Southworth-Anthoensen Press, 1938), p. 165; Fannie Chase, *Wiscasset in Pownalborough: A History of the Shire Town and the Salient Historical Features of the Territory Between the Sheepscot and Kennebec Rivers* (2nd. ed., Wiscasset, Maine [Portland, Maine]: Anthoensen Press, 1967), p. 417.

29. Chase, *Ibid.*

30. *Ibid.*

31. Charles Boardman Hawes, *Whaling* (Garden City, New York: Doubleday, Page & Company, 1924), p. 151.

32. *Ibid.*; Chase, p. 417.

33. Tower, p. 128.

34. Rice, p. 166; see also Kenneth R. Martin, "Voyages of Maine Whaleships," *Down East: The Magazine of Maine*, XVIII, 8 (May 1972), p. 47.

35. Wood, "Ship Wiscasett [sic] of Wiscasett."

36. "Wiscasset of Wiscasset" (incised drawing on a sperm whale's tooth, 1836), Kendall Whaling Museum, Sharon, Massachusetts.

37. Rice, p. 167.

38. A Journal of a Voige [sic] to the Pacific Ocean Ship C. Mitchell (manuscript journal, 1834–1837, no. 50, Kendall Whaling Museum, Sharon, Massachusetts), 2 April 1836.

39. Chase, p. 418.

40. *Ibid.*

41. *Ibid.*

42. *Ibid.*; Hawes, p. 152.

43. Chase, p. 419.

44. Dorsett, p. 38.

45. Rice, p. 166.

46. Wood, "Ship Wiscasett [sic] of Wiscasett."

47. *Ibid.*

48. *Ibid.*; Starbuck, pp. 352–353. In addition, Hawes mentions 1,700 pounds of baleen (p. 152); Rice puts it at ten times that amount (p. 166).

49. William Hutchinson Rowe, *The Maritime History of Maine: Three Centuries of Shipbuilding and Seafaring* (1948; rpt. Freeport, Maine: Bond Wheelwright Company, n.d.), p. 270; Hawes, p. 152.

50. Chase, p. 417n.

51. Starbuck, pp. 352–353; Hawes, p. 152; Chase, p. 419.

52. Rice, p. 166; Chase, p. 419; Carl C. Cutler, *Queens of the Western Ocean: The Story of America's Mail and Passenger Sailing Lines* (Annapolis: U.S. Naval Institute Press, 1961), p. 515; Hawes, p. 153.

53. Jane Day, "Gil Whitman's Anvil Art," *Down East: The Magazine of Maine*, XVIII, 7 (April 1972), pp. 49–50.

54. Densmore, *op. cit.*

55. An article by Kenneth R. Martin, "Spirit Voices and a Bath Whaler," covering in different form and detail David Densmore's whaling career, appears in *Down East: The Magazine of Maine*, XX, 4 (November 1973), pp. 42–45, 65. Hereafter cited as "Spirit Voices."

56. Densmore, pp. 99, 117, 120–122, or see Kenneth Martin, "Spirit Voices," p. 43.

57. Densmore, p. 139.

58. *Ibid.*, p. 142.

59. Bath District Customs Records (Robert B. Applebee Collection, Stockton Springs, Maine). These records indicate a possible lapse of memory on Densmore's part: the name of James F. Patten, not John Patten, appears as one of the vessel's owners.

60. Densmore, pp. 143–144.

61. *Ibid.*, pp. 145–146. Densmore claims the vessel was held up until November, but other records show she sailed on 19 October 1841 (Wood, "Bark Massasoit of Bath").

62. Densmore, p. 151.

63. *Ibid.*, p. 159.

64. Wood, "Bark Massasoit of Bath."

65. *Ibid.*

66. *Ibid.*

67. *Ibid.*, Densmore, pp. 163, 168.

68. Densmore, p. 163.

69. *Ibid.*, p. 179.

70. William Torrey Notarial Records (manuscript records, Bath Marine Museum, Bath, Maine) vol. 6 (1835–1846), pp. 157–158; Densmore, p. 180.

71. Densmore, pp. 182–183.

72. Bath Customs Records (Robert B. Applebee Collection).

73. Starbuck, pp. 484–485.

74. *Eastern Argus*, 2 June 1841.

75. Starbuck, pp. 386–387.

76. Wood, "Schr. Warwick of Bucksport (Me.)."

77. *Ibid.*

78. *Ibid.*; Starbuck, pp. 386–387.

79. Wood, "Brig Beulah of Bangor."

80. George F. Neil, Journal kept aboard the Ship *Elbe* (manuscript journal, 1845–1848, Barbara E. Johnson Whaling Collection, Princeton, New Jersey).

81. Register, Ship *Lucy Ann*, Wilmington, 20 July 1839 (Records of the Bureau of Marine Inspection and Navigation [Record Group 41], National Archives Building, Washington, D.C.).

82. Kenneth R. Martin, "The Successful Whaling Voyage of the *Lucy Ann* of Wilmington, 1837–1839," *Delaware History*, XV, 2 (October 1972), p. 85.

83. Starbuck, pp. 450–451.

84. Maine Whalers (typescript list compiled by Nathan Lipfert, Bath Marine Museum, Bath, Maine); George Francis Dow, *Whale Ships and Whaling During Three Centuries, With an Account of the Whale Fishery in Colonial New England* (Salem, Massachusetts: Marine Research Society, 1925), p. 309.

85. Reginald B. Hegarty, ed., *Returns of Whaling Vessels Sailing from American Ports, A Continuation of Alexander Starbuck's "History of the American Whale Fishery," 1876–1928*, Additions by Philip F. Purrington (New Bedford: Old Dartmouth Historical Society and Whaling Museum, 1959), p. 38.

86. Hohman, pp. 189–190.

87. *Ibid.*, p. 190.

88. Bath Customs Records (Robert B. Applebee Collection).

89. Maine Whalers; Mr. Charles F. Batchelder to the author (letter, Boston, 3 April 1972).

90. Maine Whalers.

91. Reports of Survey, New England

65 Shipbuilding Company, 1886 (Bath Marine Museum, Bath, Maine).

92. Hegarty, pp. 11, 31.

93. *Ibid.*, pp. 17, 23, 31, 39, 40.

94. Scammon, p. 58.

95. *Ibid.*, p. 62.

96. Lloyd C. M. Hare, *Salted Tories: The Story of the Whaling Fleets of San Francisco* (Mystic, Connecticut: Marine Historical Association, Inc., 1960), pp. 60–61.

97. Dorsett, p. 39.

98. *Ibid.*

99. Hegarty, p. 9.

100. Hare, p. 61.

101. Classified Index of Maine-built Vessels (Bath Marine Museum, Bath, Maine).

102. John R. Spears, *The Story of the New England Whalers* (New York: Macmillan Company, 1908), p. 240.

103. Hegarty, p. 9.

104. Hare, pp. 75–76.

105. Classified Index of Maine-built Vessels.

106. Captain Ellsworth Luce West, as told to Eleanor R. Mayhew, *Captain's Papers: A Log of Whaling and Other Sea Experiences* (Barre, Massachusetts: Barre Publishers, 1965), pp. 17–18. For details of the William Baylies' construction see Kenneth R. Martin, "The Bath-built Steam Whalers," *Down East: The Magazine of Maine*, XIX, 5 (January 1973), p. 83.

107. Hegarty, p. 26.

108. Reports of Survey, Kelley, Spear and Company, 1886 (Bath Marine Museum, Bath, Maine).

109. West and Mayhew, p. 28.

110. Reports of Survey, Kelley, Spear and Company, 1892 (Bath Marine Museum, Bath, Maine).

111. Hegarty, p. 11.

112. Hare, p. 110.

113. Hegarty, p. 23.

114. *Ibid.*, p. 26.

115. Robert R. Newell, *Steam Whaler Navarch Lost in the Arctic Ice* (Norwalk, Connecticut: Whale House Press, 1969), p. 3.

116. Captain George F. Tilton, *Cap'n George Fred Himself* (Garden City, New York: Doubleday, Doran and Company, Inc., 1928), pp. 166–168; Hawes, pp. 270–271.

117. *Report of the Cruise of the U.S. Revenue Cutter Bear and the Overland Expedition for the Relief of the Whalers in the Arctic Ocean* etc. (Washington, D.C.: Government Printing Office, 1899), pp. 79–93.

118. Classified File of Maine-Built Vessels; Vincent Short and Edwin Sears, *Sail and Steam along the Maine Coast* (Portland, Maine: Bond Wheelwright Company, 1959), pp. 58–59.

119. Captain Hartson Bodfish, as told to Joseph C. Allen, *Chasing the Bowhead* (Cambridge, Massachusetts: Harvard Uni-

versity Press, 1936), pp. 12–14.

120. *Ibid.*, pp. 254–256, 259.

121. Jim Gibbs, *Pacific Square Riggers* (Seattle: Superior Publishing Company, 1969), p. 60; Hare, p. 110.

122. Gibbs, p. 61.

123. *Ibid.*, p. 157; Hare, p. 106.

124. Mary Chipman Lawrence, *The Captain's Best Mate: The Journal of Mary Chipman Lawrence on the Whaler Addison 1856–1860*, ed. by Stanton Garner (Providence: Brown University Press, 1966), pp. 254, 262.

125. J. Ross Browne, *Etchings of a Whaling Cruise* (1846; rpt. ed. by John Seelye, Cambridge, Massachusetts: Belknap Press of Harvard University Press, 1968), *passim.*

126. Chester Howland, *Thar She Blows!* (New York: Wilfred Funk, 1951), p. 168.

127. Philip F. Purrington, *4 Years A-whaling: Charles S. Raleigh, Illustrator* (New Bedford, Massachusetts [Barre, Massachusetts]: Published for the Whaling Museum, by Barre Publishers 1972), p. 13.

128. *Eastern Argus*, 2 June 1845; see also Kenneth R. Martin, "The Folk Art of Maine's Whalemen," *The Quarterdeck*, II, 11 (Spring–Summer 1972), p. 10.

129. *Tri-Weekly Argus* (Portland, Maine), 3 September 1845.

130. *Eastern Argus*, 17 March 1847.

131. *Ibid.*, 23 November 1852; see also Kenneth R. Martin, "The Folk Art of Maine's Whalemen," p. 10.

132. "Remarks on board Ship Eliza Adams of New Bedford" (single page of anonymous entries, October 1850), found on a back page of Weston M. Cornell, Journal kept aboard Bark *Rhine* of New Bedford, 1846–1848 (manuscript journal, no. 394, Kendall Whaling Museum, Sharon, Massachusetts).

BIBLIOGRAPHY

UNPUBLISHED SOURCES

(In the following list, the manuscript writers' own titles, if known, have been preserved.)

Bath Customs District Records. Robert B. Applebee Collection, Stockton Springs, Maine.

Bonney, Henry M. Remarks aboard the Ship Lucy Ann of Wilmington, Delaware, 1837–1839. Manuscript journal, The Whaling Museum, New Bedford, Massachusetts.

Classified Index of Maine-built Vessels. Bath Marine Museum, Bath, Maine.

Cornell, Weston M. Journal kept aboard Bark *Rhine* of New Bedford, 1846–1848. Manuscript journal, no. 394, Kendall Whaling Museum, Sharon, Massachusetts.

For my Own Gratification & Satisfaction I Write or Endeavour to Write a Journal or an Slightly Sketch of the Most Noted Proceedings & Transactions Events & Observations Remarks & Occurrences . . . which generally Arises on a Voyage a-Whaling to the Pacific. Manuscript journal kept aboard Ship *Canton,* 1834–1838, no. 43, Kendall Whaling Museum, Sharon, Massachusetts.

Gifford, Phylander. Remarks on Board Ship Addison. Manuscript journal, 1838–1841, no. 5, Kendall Whaling Museum, Sharon, Massachusetts.

Journal kept aboard Ship *Canton* of New Bedford, 1835–1838. Manuscript journal, no. 252, Kendall Whaling Museum, Sharon, Massachusetts.

Journal kept aboard Ship *Parachute* of New Bedford, 1838–1840. Manuscript journal, no. 165, Kendall Whaling Museum, Sharon, Massachusetts.

A Journal of a Voige to the Pacific Ocean Ship C[hristopher] Mitchell. Manuscript journal, 1834–1837, no. 50, Kendall Whaling Museum, Sharon, Massachusetts.

Maine Whalers. Typescript list compiled by Nathan Lipfert. Bath Marine Museum, Bath, Maine.

Martin, John F. Journal of a Voyage in the Whale Ship Lucy Ann of Wilmington, Del., 1841–1844. Manuscript journal, Chicago Historical Society Library, Chicago, Illinois.

Neil, George F. Journal kept aboard the Ship *Elbe,* 1845–1848. Manuscript journal, Barbara E. Johnson Whaling Collection, Princeton, New Jersey.

Norton, Charles C. Journal of a Voyage to the Pacific Ocean in the Ship Canton.

Manuscript journal, 1834–1838, no. 355, Kendall Whaling Museum, Sharon, Massachusetts.

Register, Ship *Lucy Ann*, Wilmington, 20 July 1839, Records of the Bureau of Marine Inspection and Navigation (Record group 41), National Archives Building, Washington, D.C.

Reports of Survey Records, New England Shipbuilding Company, 1886. Bath Marine Museum, Bath, Maine.

Reports of Survey Records, Kelley, Spear and Company, 1888 and 1892. Bath Marine Museum, Bath, Maine.

Snell, Angles. Journal of a Whaling Voyage to the Atlantic Ocean by Angles Snell on board Brig Ann Maria Fall River. Manuscript journal, 1837–1838, no. 17, Kendall Whaling Museum, Sharon, Massachusetts.

Torrey, William. Notarial Records. Bath Marine Museum, Bath, Maine.

Wood, Dennis. Abstracts of Whaling Voyages from the United States, 1831–1873. Manuscript abstracts, Melville Whaling Room, Free Public Library, New Bedford, Massachusetts.

PUBLISHED SOURCES

American Shipmasters' Association. *Record of American and Foreign Shipping: 1878*. New York: American Bureau of Shipping, 1878.

Ashley, Clifford. *The Yankee Whaler*. Boston: Houghton Mifflin Company, 1926.

Bodfish, Captain Hartson H., as told to Joseph C. Allen. *Chasing the Bowhead*. Cambridge, Massachusetts: Harvard University Press, 1936.

Brewington, M. V., and Dorothy. *Kendall Whaling Museum Paintings*. Sharon, Massachusetts: Kendall Whaling Museum, 1965.

Browne, J. Ross. *Etchings of a Whaling Cruise*. 1846; rpt. edited by John Seelye. Cambridge, Massachusetts: Belknap Press of Harvard University Press, 1968.

Bulletin of the United States Fish Commission, Volume V, for 1885. Washington, D.C.: Government Printing Office, 1886.

Burrage, Henry S., ed. *Early English and French Voyages, Chiefly from Hakluyt*. New York: Charles Scribner's Sons, 1906.

Calderwood, Ivan E. *Sequel . . . Days of Uncle Dave's Fish House*. Rockland, Maine: Courier Gazette, Inc., 1972.

Camp, Helen. *Pemaquid Lost and Found*. N.p., 1967.

Chase, Fannie S. *Wiscasset in Pownalborough: A History of the Shire Town and the Salient Historical Features of the Territory Between the Sheepscot and Kennebec Rivers*. 2nd edition, Wiscasset, Maine [Portland, Maine]: Anthoensen Press, 1967.

Coffin, Robert P. Tristram. *Kennebec: Cradle of Americans.* New York and Toronto: Rinehart & Company, Inc., 1937.

Cook, Captain John A. *Pursuing the Whale: A Quarter Century of Whaling in the Arctic.* London: John Murray, 1926.

Cutler, Carl C. *Queens of the Western Ocean: The Story of America's Mail and Messenger Sailing Lines.* Annapolis: U.S. Naval Institute Press, 1961.

Daily Evening Advertiser. Portland, Maine, 1836.

Davis, William M. *Nimrod of the Sea: Or the American Whaleman.* 1874; rpt. Boston: Charles E. Lauriat Co., 1926.

Day, Jane. "Gil Whitman's Anvil Art." *Down East: The Magazine of Maine,* XVIII, 7 (April 1972), pp. 48–51.

Densmore, David C. *The Halo: An Autobiography.* Boston: Voice of Angels Publishing House, 1876.

Dorsett, E. Lee. "The Maine Whalers." *Down East: The Magazine of Maine,* II, 6 (April 1956), pp. 14–15, 38–39.

Dow, George Francis. *Whale Ships and Whaling: A Pictorial History of Whaling During Three Centuries, with an Account of the Whale Fishery in Colonial New England.* Salem, Massachusetts: Marine Research Society, 1925.

Earle, Walter K. *Scrimshaw, Folk Art of the Whalers.* Cold Spring Harbor, New York: Whaling Museum Society, Inc., 1957.

Eastern Argus. Portland, Maine, 1839–1843.

Federal Writers Project of the Works Progress Administration of the State of Massachusetts. *Whaling Masters.* American Guide Series. New Bedford: Old Dartmouth Historical Society, 1938.

Gibbs, Jim. *Pacific Square Riggers.* Seattle: Superior Publishing Company, 1969.

Goode G. Browne, et al., *The Fisheries and Fishery Industries of the United States: Section V, History and Methods of the Fisheries.* 3 vols., Washington, D.C.: Government Printing Office, 1887.

Hare, Lloyd C. M. *Salted Tories: The Story of the Whaling Fleets of San Francisco.* Mystic, Connecticut: Marine Historical Association, Incorporated, 1960.

Hawes, Charles Boardman. *Whaling.* Garden City, New York: Doubleday Page & Company, 1924.

Hegarty, Reginald B., ed. *Addendum to "Starbuck" and "Whaling Masters."* New Bedford: New Bedford Free Public Library, 1964.

Hegarty, Reginald B., ed. *Returns of Whaling Vessels Sailing From American Ports, a Continuation of Alexander Starbuck's "History of the American*

Whale Fishery," 1876–1928. Additions by Philip F. Purrington. New Bedford: Old Dartmouth Historical Society and Whaling Museum, 1959.

Hohman, Elmo Paul. *The American Whaleman*. 1928; rpt. New York: Library Editions, Ltd., 1970.

Howland, Chester S. *Thar She Blows!* New York: Wilfred Funk, Inc., 1951.

Lawrence, Mary C. *The Captain's Best Mate: The Journal of Mary Chipman Lawrence on the Whaler Addison*, 1856–1860. Edited by Stanton Garner. Providence: Brown University Press, 1966.

McCormick, Edgar, and Edward McGehee, eds. *Life on a Whaler: Selected Source Materials for College Research Papers*. Boston. D. C. Heath and Company, 1960.

Martin, Kenneth R. "Spirit Voices and a Bath Whaler." *Down East: The Magazine of Maine*, XX, 4 (November 1973), pp. 42–45, 65.

Martin, Kenneth R. "The Successful Whaling Voyage of the *Lucy Ann* of Wilmington, 1837–1839." *Delaware History*, XV, 2 (October 1972), pp. 85–103.

Martin, Kenneth R. "Voyages of Maine Whaleships." *Down East: The Magazine of Maine*, XVIII, 8 (May 1972), pp. 46–49 and *passim*.

Matthews, Leonard H., *et al. The Whale*. New York: Simon and Schuster, 1968.

Morison, Samuel Eliot. *The Maritime History of Massachusetts, 1783–1860*. Boston: Houghton Mifflin Company, 1921.

National Archives Project. *Ship Registers of New Bedford, Massachusetts*. Compiled by The Survey of Federal Archives, Division of Professional and Service Projects, Work Projects Administration: 3 vols., Boston: National Archives Project, 1940.

Newell, Robert R. *Steam Whaler Navarch Lost in the Arctic Ice*. Norwalk, Connecticut: Whale House Press, 1969.

Owen, Henry W. *The Edward Clarence Plummer History of Bath, Maine*. Bath, Maine: Times Company, 1936.

Photographs of Whaling Vessels. Prepared by Erik A. R. Ronnberg. A Pictorial Supplement of *The American Neptune*. Issued Separately as Old Dartmouth Historical Sketch Number 77. New Bedford: Old Dartmouth Historical Society, n.d.

Proper, Ida S. *Monhegan, the Cradle of New England*. Portland, Maine: Southworth Press, 1930.

Purrington, Philip F. *4 Years A-whaling: Charles S. Raleigh, Illustrator*. Barre, Massachusetts: Published for the Whaling Museum, New Bedford, Massachusetts, by Barre Publishers, 1972.

Report of the Cruise of the U.S. Revenue Cutter Bear and the Overland Expedition for the Relief of the Whalers in the Arctic Ocean, etc. Washington, D.C.: Government Printing Office, 1899.

Rice, George W. *The Shipping Days of Old Boothbay from the Revolution to the World Wr with Mention of Adjacent Towns.* Boothbay Harbor, Maine [Portland, Maine]: Southworth-Anthoensen Press, 1938.

Rowe, William Hutchinson. *The Maritime History of Maine: Three Centuries of Shipbuilding and Seafaring.* 1948; rpt. Freeport, Maine: Bond Wheelwright Company, n.d.

Scammon, Charles M. *The Marine Mammals of the North-western Coast of North America: Together with an Account of the American Whale Fishery.* 1874; rpt. New York: Dover Publications, Inc., 1968.

Sherman, Stuart C. *The Voice of the Whaleman with an Account of the Nicholson Whaling Collection.* Providence: Providence Public Library, 1965.

Short, Vincent, and Edwin Sears. *Sail and Steam along the Maine Coast.* Portland, Maine: Bond Wheelwright Company, 1955.

Smith, Captain John. *Travel and Works of Captain John Smith, President of Virginia and Admiral of New England,* ed. by Edward Arber. 2 vols., Edinburgh: John Grant, 1910.

Spears, John R. *The Story of the New England Whalers.* New York: Macmillan Company, 1908.

Stackpole, Edouard. *Scrimshaw at Mystic Seaport.* Mystic, Connecticut: Marine Historical Association, Incorporated, 1958.

Starbuck, Alexander. *History of the American Whale Fishery from its Earliest Inception to the Year 1876.* 1878; rpt. 2 vols., New York: Argosy-Antiquarian Ltd., 1964.

Tilton, George F. *Cap'n George Fred Himself.* Garden City, New York: Doubleday, Doran and Company, Inc., 1928.

Tower, Walter S. *A History of the American Whale Fishery.* University of Pennsylvania Publications on Political Economy and Public Law, no. 20. Philadelphia: Published for the University of Pennsylvania, 1907.

Tri-Weekly Argus. Portland, Maine, 1845–1847.

Verrill, A. Hyatt. *The Real Story of the Whaler: Whaling, Past and Present* 1916; rpt. New York: D. Appleton and Company, 1931.

West, Ellsworth Luce, as told to Eleanor R. Mayhew. *Captain's Papers: A Log of Whaling and Other Sea Experiences.* Barre, Massachusetts: Barre Publishers, 1965.

Winslow, Helen. *The Folk Art of the American Whaleman.* Nantucket: Nantucket Historical Association, 1965.